COW
Mummy – **Cow**
Daddy – **Bull**
Baby – **Calf**

GOOSE
Daddy – **Gander**
Mummy – **Goose**
Baby – **Gosling**

CHICKEN
Mummy – **Hen**
Daddy – **Cockerel**
Baby – **Chick**

Adam's FARM

Adam Henson

Illustrated by
Hannah Abbo

PUFFIN

Hello! I'm **Adam** and I'm a farmer
(I'm a TV presenter too, so you might recognize me).

I've been farming all my life. I grew up on a farm and my dad was a farmer, so as soon as I could carry a bucket of water or bottle-feed a lamb, I was helping him out. I love animals and being in the countryside, and I always knew that I wanted to follow in my dad's footsteps.

Today, I look after the same farm I grew up on. It's 650 hectares – which is about the size of a thousand football pitches – and we have around 2,000 animals. There are cattle, sheep, goats, pigs, horses, rabbits, chickens, ducks and geese (plus a few dogs).

I think farming is the best job in the world, and I get asked questions about it all the time, so I wanted to answer some of these for you. Some questions have been sent to me for this book (you will see the names and ages of the children under their questions) and some are my favourite questions that I've been asked by children visiting my farm.

But first, I have a question for **YOU**.

Are you ready to learn about everything that happens on a farm? From the busy life of a farmer to the animals big and small; from the powerful machines to how your food makes its way from my farm to your plate.

Yes! Let's get going then.

WHAT WAS IT LIKE GROWING UP ON A FARM?

It was **a lot** of fun. Me and my three older sisters were always spending time around the animals. We had pet rabbits and lambs. There were lots of chickens and their chicks running around the farmyard too, and cockerels crowing in the morning. We even had a pet reindeer called Rudolph (of course!) and he used to pull us along on a sledge when it snowed.

We'd also go out picking blackberries and apples for my mum to make delicious pies with.

DID ANY OF THE FARM ANIMALS LIVE IN THE HOUSE?

They did! If there was a poorly lamb, we'd bring it inside, place it in a basket by the cooker to warm it up and bottle-feed it until it was healthy again. My favourite lamb was Friendly Fred. He lived in the house when he was little. He would come for walks with us and the dogs, and he believed he was a dog too! If he saw a fox in the fields, he would run towards it thinking it was another dog, but the bell round his neck would frighten the fox away.

WHAT TIME DO YOU WAKE UP TO START YOUR JOB AND FEED THE ANIMALS?

(Chloe, 10)

We usually start work on the farm at eight in the morning. But in March and April, when it's the peak of the lambing season, we start early and finish late. Lambing season is the time of year when sheep give birth to their lambs. Sometimes we have to get up at two in the morning and work all night to make sure the animals are healthy!

Dairy farmers – who keep cows – will often milk their cows at 4.30 in the morning and then again twelve hours later at 4.30 in the afternoon.

IS THERE ANY TIME FOR 'YOU' TIME WHEN WORKING ON THE FARM?

(Katrina, 11)

There's **always** something to do, but I'm lucky that I live on the farm, so I'm often popping into the house to see my family and have a cup of tea. I relax in the evenings, but even when I'm working outdoors, I'm happy and enjoying myself!

WHAT ARE THE DAILY JOBS YOU DO ON THE FARM?

(Tyler, 10)

All the animals need to be checked on at least once every day. We call it 'checking round'. We start in the farmyard with the pigs, geese and ducks. Then we head out to the fields with our sheepdog, Gwen, to see the cattle and sheep. (A sheepdog is a dog that helps farmers move sheep around fields.) We check that the animals are safe and that they have plenty of food and water. We also do jobs such as fixing fences and machinery.

In early summer, we cut the sheep's wool to keep them cool – this is called shearing.

In the winter, when the cattle are keeping warm in the sheds, we make sure they have plenty of clean, dry straw to lie on.

On our farm, we plant seeds in the autumn and spring, then we protect these crops from diseases and spread fertilizer to help them grow. We harvest them in the summer.

HOW DO YOU KNOW WHEN YOUR ANIMALS ARE HAPPY OR SAD?

(Harry, 4)

Often if a cow is lying down and munching on food, it means they are relaxed and calm. With lambs, calves, foals and goat kids, if you see them playing, you know they are happy and having fun.

But if their ears are down and they're a bit hunched up, it's usually because they are ill. They might stay in the corner of a field and not eat their food. We look after poorly animals and get them back to good health!

ARE THERE ANY UNLIKELY FRIENDS ON THE FARM?

The most unlikely friendship on the farm was between a cow and some piglets. We used to have a special breed of cattle known as Dexter cattle. They are very short, which means their udders (the part that makes milk) are lower to the ground. We had some piglets who would drink milk from one of the Dexter cows! She didn't mind though, and the piglets enjoyed the extra milk.

Iron-age Piglets

We also have a duck called Desmond and an Indian Game cockerel called Frank who are best friends!

WHAT IS A RARE BREED?

This is the name for a breed of animals that is no longer found on most farms and so there aren't many of them left. Today we have **fifty rare breeds** on our farm, including Soay sheep – which are a small brown breed of sheep – and White Park cattle, which have black patches round their eyes and on their ears, noses and knees.

Can you spot the White Park cattle?

WHAT DO THE DIFFERENT ANIMALS GET FED?

(Jake, 11)

We give pigs, chicken, geese and ducks special hard food called pignuts or chicken pellets every day. These are mixes of grains and include everything they need for a healthy diet.

The cattle, sheep and goats either eat grass that has grown in their field, or hay, which is grass that has been cut and dried in the summer and stored to feed them in the winter months.

Golden Guernsey Goat

WHICH ANIMAL HAS THE SMELLIEST POO ON THE FARM?

Sheep do little pellets of poo that drop out in small balls and cattle have big, sloppy cow pats. Pig poo comes out in lumps and is very smelly. But the worst is duck poo – **it stinks!**

Animal poo is really important on a farm. We collect it and put it on a big pile called a muck heap. Tiny living things make their way inside the heap and break down the poo, which is full of lots of nutrients. Then we spread it across the fields, where it acts as a brilliant fertilizer, helping our crops and grass to grow.

DO ANIMALS EVER ESCAPE?

(Leo, 3)

Yes, they do! If a gate is left open by mistake or if a tree has fallen on a fence and broken it, they tend to find a way out. We have to get them back in their field quickly because it is easy for them to wander off and get lost.

We have an old Shetland pony called Dougal who is very clever at opening gate latches, so in his field we have **to chain them shut!**

WHAT ARE THE NAMES OF THE DIFFERENT MACHINES AND EQUIPMENT, AND WHAT ARE THEY USED FOR?

(Chloe, 10)

The tractor is the **most important** machine on the farm. It pulls lots of other types of equipment that each have their own jobs. It pulls the plough, which turns over the soil to bury the weeds (plants that that grow in places they are not wanted). It also pulls the cultivator, which smooths the soil to help make the field ready for planting seeds. It then pulls the drill, which plants the seeds in the soil, and then the set of heavy rolls that press the seeds down.

A tractor can also tow a trailer to help us move things around the farm, and a baler that collects straw or grass and cleverly makes it into bundles called bales, which are fed to the animals in the winter.

The combine harvester is a very large, smart machine with lots of jobs to do. It cuts the crops – such as wheat – when they are ready. It then separates the grain from the straw, and collects the grain in its large tank – grain is the part that can be eaten by people. Next, the straw falls out of the back in rows, ready to be made into bales.

WHAT IS THE MOST EXPENSIVE MACHINE ON THE FARM?

A big combine harvester can cost around £500,000. You could buy a house with that much money!

WHY DO FARMERS USE MACHINES INSTEAD OF THEIR HANDS TO MILK COWS?

(Arthur, 6)

Cows used to be milked by hand, but machines make it much quicker. A person might be able to milk fifteen cows a day by hand, while machines can milk hundreds! There are now robots that can milk up to fifty cows in a day, working all day and night.

HOW BIG ARE TRACTORS?

(Ora, 4)

Tractors range from tiny ones that could fit in a small garden, to massive machines that pull **gigantic** farming equipment. The size of a tractor is measured in something called horsepower. A little tractor might only be 30 horsepower, which means it can do the same amount of work as around 30 horses. However, a big one might be 400 horsepower!

WHAT TRACTORS DO YOU HAVE ON YOUR FARM?

We have different sized tractors for different jobs. There is a little orange one called a Kubota, which pulls small trailers around the farm, and the forage wagon, which spreads the animal feed. Then we've got a bright yellow JCB loader. It's a great machine that has a bucket for carrying grain, a spike for moving hay bales and something called a post rammer for building fences. And there's the big green John Deere tractor. The larger ones have tracks instead of wheels, which can stop them from damaging soil as they drive across fields, pulling the cultivator or drill.

And we also have a combine harvester, of course!

I LOVE TRACTORS! WHAT AGE CAN I DRIVE A TRACTOR ON A FARM?

(Hugo, 3)

Hugo has a few years to wait yet. You can drive a tractor with low horsepower on a farm when you are thirteen. This is four years earlier than you can drive a car on a road in the UK.

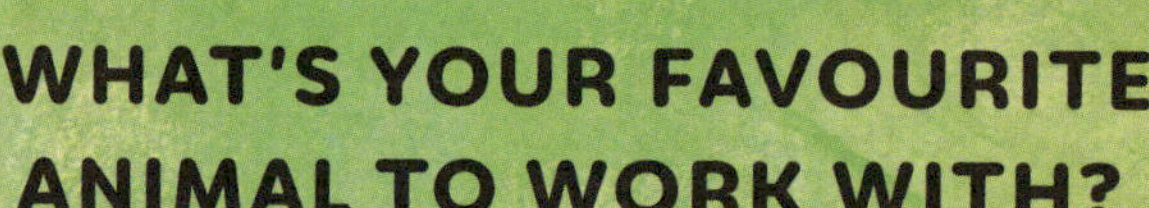

WHAT'S YOUR FAVOURITE ANIMAL TO WORK WITH?

(Riley, 10)

I love my dogs. I've got three pet dogs in the house – Mini, Hilda and Olive – and two Border collies called Peg and Gwen, who are working sheepdogs. Peg is too old to work now, but Gwen still helps me move the sheep. She's very clever and fun to work with.

HOW DO YOU TRAIN YOUR SHEEPDOGS?

The first thing you have to do is teach sheepdogs to come back to you when you ask them. Then you introduce them to the sheep and teach them special commands. '**Away**' means they need to run to the right of the sheep; '**come-by**' means they need to run to the left; '**walk on**' means they need to get up and start walking; and '**stop**' means stop! Once they've understood this, they can help you move the sheep wherever you want them to go.

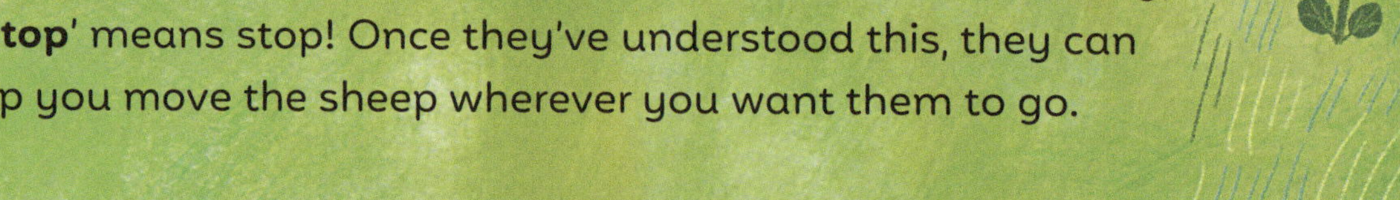

WHAT HAPPENS IF YOU DON'T GIVE A SHEEP A HAIRCUT?

Many sheep never stop growing wool. So, if you don't cut their fleece, they will grow another fleece on top of the one they already have. This can be dangerous as it will make them very hot. It's the law in the UK that a sheep's wool must be cut – or shorn – once a year.

Some ancient breeds of sheep, such as the Soay, have wool that starts to fall out (or shed) naturally in the springtime when the weather warms up.

DO COWS ACTUALLY HAVE FOUR STOMACHS AND WHY?

(Arlo, 5)

Arlo is right – cows do have four stomachs to help them break down (or digest) all the grass they eat. One of their stomachs is called the rumen and it is full of tiny living things that break down the grass. Cattle will chew their food, swallow it, then bring it back up to their mouth to chew again! This is called chewing the cud. Sheep, goats and deer have four stomachs too. These animals are called ruminants.

WHY DO HIGHLAND COWS HAVE BIG HORNS?

(Arthur, 8)

It's only the females that have big horns. They're a very old breed of cattle, and in the past they used their horns to protect their calves from wolves. The males – or bulls – also have horns, but they're much thicker and shorter.

WHY DO COWS LIE DOWN WHEN IT'S RAINING?

(Theo, 8)

Many people believe that cattle lie down when it's about to rain, but actually this isn't true. They lie down for all sorts of reasons – to relax, chew their food or sleep.

HOW LONG IS A COW PREGNANT FOR?

(George, 7)

Cows are pregnant for nine months, which is the same length of time as humans. Sheep and goats are pregnant for five months, and for pigs it's three months, three weeks and three days. A chick grows in the egg for twenty-one days and a duckling for twenty-eight days before hatching.

WHY DO SOME COWS HAVE THEIR HORNS UP AND WHY DO SOME HAVE THEM DOWN?

(Ronnie, 8)

There's no real reason. Even within the same breed of cattle, their horns can look different. For example, some Longhorn cattle have upright horns and others have curled horns, known as bonnet horns.

WHAT THINGS CAN YOU MAKE FROM MILK?

You can make **lots** of things from milk, such as butter, cheese, ice cream and yogurt. I love a bowl of cereal and milk in the morning!

My favourite flavour is strawberry milk.

What's yours?

OTHER THAN COWS, WHICH ANIMALS MAKE MILK?

Mammals are animals that make milk to feed their young. Dogs and cows are mammals, and so are humans and whales! Many farmers milk cows, sheep and goats and in some parts of the world, people milk horses and camels too.

DOES CHOCOLATE MILK COME FROM CHOCOLATE COWS?

(Talah, 6)

This is a great question from Talah, but, no, there are no chocolate cows. You have to add an ingredient called cocoa to milk to make it chocolatey. You can make **all sorts** of flavours, such as strawberry, banana and raspberry – it just depends what you add to the milk!

HOW DOES MILK GET FROM A COW TO MY CEREAL BOWL?

Cows are milked twice a day. They have teats under their body where the milk comes from. Something called a cluster goes on to the cow's teats and squeezes the milk out.

The milk goes from the cow, down the pipes and into a big metal tank, which keeps it nice and cool.

Every day – sometimes twice a day – the milk is taken away to a place called a dairy.

Here it is either filtered to make sure it's nice and clean and put into milk bottles, or it is turned into something else, such as cheese or yogurt. Then it is sent off to shops for you to buy.

It only takes one or two days for milk to travel from the cow to your fridge.

WHAT DO YOU GROW ON THE FARM?

We grow lots of crops. There is wheat, which is turned into flour for baking bread and cakes. There are peas and beans that we use in animal feed. And there's rye that goes into rye bread. We also grow a plant called oilseed rape. It has yellow flowers, which make oil that we use in cooking.

WHICH ANIMAL DOES BEEF COME FROM?

Beef comes from cattle. On our farm we keep cattle for beef, sheep for lamb (and wool) and pigs for pork sausages and bacon. Some farmers keep cows for milk, and some raise chickens, ducks and turkeys, called poultry, for meat and eggs.

HOW DOES WHEAT BECOME BREAD?

Bread is made from flour, which is normally made from the wheat crop. Wheat takes nearly one year to grow, so we plant it in October and harvest it the following August.

We then store it in our sheds until we sell it. It is usually bought by the miller, who takes it away in a great big lorry and grinds the wheat down to make flour. That flour is then sold to the baker.

And finally, the baker adds yeast, water and other ingredients to make bread. Then they sell it to the shops and supermarkets where you buy it. It takes about a year and a half from planting wheat to the bread being in your tummy.

WHEN I GROW UP, I REALLY WANT TO BE A SHEEP FARMER AND HAVE LOTS OF SHEEP. WHAT ADVICE DO YOU HAVE TO HELP ME WHEN I'M OLDER?

(Eleanor, 6)

Being a sheep farmer is a **great** job. Eleanor – and anyone else who also wants to be a sheep farmer – will need to be hard-working, enjoy being outdoors and love animals! You will also need to go to college or university after you finish school to learn all about looking after sheep. There you will learn about what to feed them, the medicines they might need and much more. And then, when you are old enough, you could try and help some sheep farmers on their farm before you get your own flock.

Farming shows are a great place to see lots of different breeds and find out more about them.

I'M NOT FROM A FARMING FAMILY. CAN I GET INTO FARMING?

Absolutely! Farming is a job open to everyone. Once you're ten, you can join a Young Farmers' Club. There are clubs all over the UK, and they are very welcoming to people who aren't farmers. It's a great place to start your journey. When you're older, you can also go to university to learn all about the different types of farming.

WHAT'S THE MOST CHALLENGING THING IN FARMING TODAY?

(Lillie, 9)

The biggest challenge hasn't changed much over the years – the weather! You need lovely, dry conditions to plant crops and you need rain to help them grow, but it doesn't always happen that way. If it's too frosty or windy during the winter months, the crops can become damaged. And if there's not enough rain or there's so much that it floods, the crops don't grow very well. You can be the best farmer in the world, **but you can't beat the weather!**

WHAT IS THE MOST EXCITING THING ABOUT THE FUTURE OF FARMING?

All the smart technology. There are self-steering tractors, like my John Deere tractor, which has a computer on board and clever electronics to help it drive all by itself! And there are devices that tell you when animals are about to give birth. You can also now use robots to pick strawberries or to scrape up all the animal poo and even give a cow a back-scratch. And there are drones that find and kill harmful weeds that damage crops.

HOW HAS FARMING CHANGED SINCE YOU WERE YOUNG?

When I was a child, you mainly needed to be strong and love the outdoors to be a farmer. But now being good with technology is a big part of it too. If you love playing computer or video games, then you could have the right skills to become a farmer. So much of farming today is about being able to use all the clever technology that helps make the job easier!

DO YOU EVER TAKE YOUR ANIMALS ON HOLIDAY?

(Ollie, 5)

We have taken our pet dogs on holiday before. We also take the farm animals to shows, where our cattle and sheep compete for prizes against other animals. The judges look at their size, colour, horn shape, teeth and feet before they pick a winner for each breed. That's a bit of a holiday for them!

DO THE ANIMALS GET A CHRISTMAS DINNER?

Unfortunately not. Animals need to eat the same food every day so they don't get an upset tummy, which is why we can't give them extra food or treats at Christmas. We still wish them a happy Christmas though!

WHAT IS YOUR FAVOURITE ANIMAL JOKE?

What do you get if you cross a sheep with a kangaroo?

A woolly jumper

Goodbye!

Thank you for all your brilliant questions. I have loved answering them and telling you about life on a farm. Hopefully you all have the chance to visit a farm one day or to get out into the countryside – it's my favourite place to be. And I hope, for some of you, this has inspired you to think about becoming a farmer yourself one day!

I don't want your questions and curiosity to end here – ask your parents, caregivers or teachers if you have more questions. And if they don't know, why not try and find the answers together?

It's time for me to get back to my animals now – I think I can hear Desmond quacking outside!

PUFFIN BOOKS

UK | USA | Canada | Ireland | Australia | India | New Zealand | South Africa

Puffin Books is part of the Penguin Random House group of companies
whose addresses can be found at global.penguinrandomhouse.com.

www.penguin.co.uk www.puffin.co.uk www.ladybird.co.uk

First published 2024
001

Printed in Dubai

The authorized representative in the EEA is Penguin Random House Ireland,
Morrison Chambers, 32 Nassau Street, Dublin D02 YH68

A CIP catalogue record for this book is available from the British Library

ISBN: 978–0–241–66234–2

All correspondence to: Puffin Books, Penguin Random House Children's
One Embassy Gardens, 8 Viaduct Gardens, London SW11 7BW

RABBIT

Mummy – **Doe**
Daddy – **Buck**
Baby – **Kit**

Mummy – **Nanny**
Baby – **Kid**
GOAT
Daddy – **Billy**